ENERGY SECTOR STANDARD OF THE PEOPLE'S REPUBLIC OF CHINA

中华人民共和国能源行业标准

Specification for Investigation of Reservoir Special Items for Hydropower Projects

水电工程水库专项工程勘察规程

NB/T 10141-2019

Chief Development Department: China Renewable Energy Engineering Institute

Approval Department: National Energy Administration of the People's Republic of China

Implementation Date: October 1, 2019

China Water & Power Press

中国水利水电出版社

Beijing 2024

All rights reserved. No part of this publication may be reproduced, stored in a retrieval system, or transmitted in any form or by any means—electronic, mechanical, photocopying, recording or otherwise, without prior written permission of the publisher.

图书在版编目（CIP）数据

水电工程水库专项工程勘察规程：NB/T 10141-2019 = Specification for Investigation of Reservoir Special Items for Hydropower Projects (NB/T 10141 -2019)：英文 / 国家能源局发布. -- 北京：中国水利水电出版社, 2024. 10. -- ISBN 978-7-5226-2760-1

Ⅰ. TV62-65

中国国家版本馆CIP数据核字第2024DV7048号

ENERGY SECTOR STANDARD
OF THE PEOPLE'S REPUBLIC OF CHINA
中华人民共和国能源行业标准

Specification for Investigation of Reservoir Special Items
for Hydropower Projects
水电工程水库专项工程勘察规程
NB/T 10141-2019
（英文版）

Issued by National Energy Administration of the People's Republic of China
国家能源局　发布
Translation organized by China Renewable Energy Engineering Institute
水电水利规划设计总院　组织翻译
Published by China Water & Power Press
中国水利水电出版社　出版发行
　　Tel: (+ 86 10) 68545888　68545874
　　sales@mwr.gov.cn
　　Account name: China Water & Power Press
　　Address: No.1, Yuyuantan Nanlu, Haidian District, Beijing 100038, China
　　http: //www.waterpub.com.cn
中国水利水电出版社微机排版中心　排版
北京中献拓方科技发展有限公司　印刷
184mm×260mm　16开本　1.75印张　55千字
2024年10月第1版　2024年10月第1次印刷

Price（定价）：￥280.00

Introduction

This English version is one of China's energy sector standard series in English. Its translation was organized by China Renewable Energy Engineering Institute authorized by National Energy Administration of the People's Republic of China in compliance with relevant procedures and stipulations. This English version was issued by National Energy Administration of the People's Republic of China in Announcement [2023] No. 5 dated October 11, 2023.

This version was translated from the Chinese Standard NB/T 10141-2019, *Specification for Investigation of Reservoir Special Items for Hydropower Projects*, published by China Water & Power Press. The copyright is reserved by National Energy Administration of the People's Republic of China. In the event of any discrepancy in the implementation, the Chinese version shall prevail.

Many thanks go to the staff from the relevant standard development organizations and those who have provided generous assistance in the translation and review process.

For further improvement of the English version, any comments and suggestions are welcome and should be addressed to:

China Renewable Energy Engineering Institute
No. 2 Beixiaojie, Liupukang, Xicheng District, Beijing 100120, China
Website: www.creei.cn

Translating organization:

POWERCHINA Chengdu Engineering Corporation Limited

Translating staff:

ZHAO Cheng	ZHANG Yixi	ZHOU Yifei	CHEN Weidong
PENG Shixiong	MA Jin'gen	WU Zhanglei	XIAO Huabo
LI Qingchun	TIAN Xiong	WANG Nengfeng	MEI Zhiping
CUI Changwu	LIU Yongbo	LUO Xiaohong	LI Changyou
SU Yu	HU Yadong	ZHANG Enming	YI Qicheng

Review panel members:

LIU Xiaofen	POWERCHINA Zhongnan Engineering Corporation Limited

YE Bin	POWERCHINA Huadong Engineering Corporation Limited
LIU Qing	POWERCHINA Northwest Engineering Corporation Limited
QI Wen	POWERCHINA Beijing Engineering Corporation Limited
PENG Peng	POWERCHINA Huadong Engineering Corporation Limited
LIU Xianggang	POWERCHINA Guiyang Engineering Corporation Limited
CHEN Li	POWERCHINA Kunming Engineering Corporation Limited
LI Shisheng	China Renewable Energy Engineering Institute
WANG Huiming	China Renewable Energy Engineering Institute

National Energy Administration of the People's Republic of China

翻译出版说明

本译本为国家能源局委托水电水利规划设计总院按照有关程序和规定，统一组织翻译的能源行业标准英文版系列译本之一。2023年10月11日，国家能源局以2023年第5号公告予以公布。

本译本是根据中国水利水电出版社出版的《水电工程水库专项工程勘察规程》NB/T 10141—2019翻译的，著作权归国家能源局所有。在使用过程中，如出现异议，以中文版为准。

本译本在翻译和审核过程中，本标准编制单位及编制组有关成员给予了积极协助。

为不断提高本译本的质量，欢迎使用者提出意见和建议，并反馈给水电水利规划设计总院。

地址：北京市西城区六铺炕北小街2号
邮编：100120
网址：www.creei.cn

本译本翻译单位：中国电建集团成都勘测设计研究院有限公司

本译本翻译人员：赵　程　张一希　周逸飞　陈卫东
　　　　　　　　彭仕雄　马金根　吴章雷　肖华波
　　　　　　　　李青春　田　雄　王能锋　梅稚平
　　　　　　　　崔长武　刘永波　罗晓红　黎昌有
　　　　　　　　粟　宇　胡亚东　张恩铭　易琪程

本译本审核人员：
刘小芬　中国电建集团中南勘测设计研究院有限公司
叶　彬　中国电建集团华东勘测设计研究院有限公司
柳　青　中国电建集团西北勘测设计研究院有限公司
齐　文　中国电建集团北京勘测设计研究院有限公司
彭　鹏　中国电建集团华东勘测设计研究院有限公司
刘祥刚　中国电建集团贵阳勘测设计研究院有限公司
陈　砺　中国电建集团昆明勘测设计研究院有限公司

李仕胜　水电水利规划设计总院
王惠明　水电水利规划设计总院

国家能源局

Announcement of National Energy Administration of the People's Republic of China [2019] No. 4

National Energy Administration of the People's Republic of China has approved and issued 297 sector standards such as *Code for Electrical Design of Photovoltaic Power Projects*, including 105 energy standards (NB), 168 electric power standards (DL), and 24 petrochemical standards (NB/SH).

Attachment: Directory of Sector Standards

National Energy Administration of the People's Republic of China

June 4, 2019

Attachment:

Directory of Sector Standards

Serial number	Standard No.	Title	Replaced standard No.	Adopted international standard No.	Approval date	Implementation date
…						
14	NB/T 10141-2019	Specification for Investigation of Reservoir Special Items for Hydropower Projects			2019-06-04	2019-10-01
…						

Foreword

According to the requirements of Document GNKJ [2015] No. 12 issued by National Energy Administration of the People's Republic of China, "Notice on Releasing the Development and Revision Plan of the Second Batch of Energy Sector Standards in 2014", and after extensive investigation and research, summarization of practical experience, consultation of relevant Chinese standards, and wide solicitation of opinions, the drafting group has prepared this specification.

The main technical contents of this specification include: general provisions, terms, basic requirements, investigation of reservoir special items, and assessment of reservoir adverse geological phenomena.

National Energy Administration of the People's Republic of China is in charge of the administration of this specification. China Renewable Energy Engineering Institute has proposed this specification and is responsible for its routine management. Energy Sector Standardization Technical Committee on Hydropower Investigation and Design is responsible for the explanation of specific technical contents. Comments and suggestions in the implementation of this specification should be addressed to:

China Renewable Energy Engineering Institute
No. 2 Beixiaojie, Liupukang, Xicheng District, Beijing 100120, China

Chief development organization:

POWERCHINA Chengdu Engineering Corporation Limited

Participating development organization:

China Three Gorges Corporation

Chief drafting staff:

PENG Shixiong	ZHANG Shishu	CHEN Weidong	MA Xingdong
XIAO Huabo	WANG Nengfeng	GU Jiangbo	YANG Zhou
QING Huabin	LIANG Yu	OU Yongsheng	WANG Kui
YIN Xianjun	XU Jun		

Review panel members:

YANG Jian	ZHANG Yijun	WANG Ruilin	ZHU Jianye
GUO Yihua	LI Wengang	JIA Yuxing	BIAN Bingqian

ZHONG Huiya	LI Xuezheng	ZHONG Guangyu	WANG Zigao
ZHAO Zhixiang	ZHANG Sihe	ZHANG Guofu	WANG Jingyong
YE Shengsheng	PIAO Ling	LI Shisheng	

Contents

1	**General Provisions**	1
2	**Terms**	2
3	**Basic Requirements**	3
4	**Investigation of Reservoir Special Items**	6
4.1	Investigation of Large- and Medium-Sized Special Items	6
4.2	Investigation of Small-Sized Special Items	8
5	**Assessment of Reservoir Adverse Geological Phenomena**	11
5.1	Landslide	11
5.2	Reservoir Bank Collapse	11
5.3	Reservoir Bank Deformation	12
5.4	Reservoir Immersion	12
5.5	Waterlogging	12
5.6	Goaf Collapse	13
Explanation of Wording in This Specification		15
List of Quoted Standards		16

1 General Provisions

1.0.1 This specification is formulated with a view to standardizing the tasks, content, methods and assessment technical requirements of investigation of reservoir special items for hydropower projects, so as to ensure the quality of investigation results.

1.0.2 This specification is applicable to the investigation of reservoir special items for hydropower projects.

1.0.3 In addition to this specification, the investigation of reservoir special items for hydropower projects shall comply with other current relevant standards of China.

2 Terms

investigation of reservoir special items

geological investigation of the special items related to land requisition and resettlement for a reservoir due to impoundment, such as railway, road, water transport, electric power and water resources facilities, which need to be constructed, renovated or extended

3 Basic Requirements

3.0.1 Geological investigation shall be arranged according to the characteristics of the reservoir special items and the requirements for the investigation tasks and level of detail at each stage, and an investigation outline shall be prepared.

3.0.2 The investigation of reservoir special items shall be divided into four stages: the prefeasibility study stage, the resettlement planning outline stage, the resettlement planning report stage, and the detailed design stage. The investigation work at each stage shall have clear targets and emphases. The contrast of investigation stages of different sectors should be in accordance with Table 3.0.2.

Table 3.0.2 Contrast of investigation stages of different sectors

Sector	Investigation stage				Remarks
Hydropower engineering	Prefeasibility study	Feasibility study		Bidding design and detailed design	–
	Prefeasibility study	Resettlement planning outline	Resettlement planning report	Detailed design	Reservoir special items
Railway engineering	Reconnaissance	Preliminary survey	Location survey	Supplementary location survey	–
Road engineering	Prefeasibility study	Project feasibility study	Preliminary design	Detailed design	–
Water transport engineering	Feasibility study		Preliminary design	Detailed design	–
Electric power engineering	Feasibility study		Preliminary design	Detailed design	–
Water resources engineering	Feasibility study		Preliminary design	Bidding design and detailed design	–

3.0.3 The level of detail of investigation for reservoir special items at each stage shall meet the following requirements:

1   At the prefeasibility study stage, the engineering geological conditions of each special item area shall be acquainted, the feasibility of the special items shall be preliminarily analyzed, the overall stability of the sites of special items shall be preliminarily assessed, and geological suggestions shall be proposed for the preliminary schemes of special items.

2   At the resettlement planning outline stage, the engineering geological conditions of each special item area shall be preliminarily ascertained, the overall stability of the sites of special items shall be assessed, and the geological data required for the comparison of alternatives and determination of schemes shall be provided.

3   At the resettlement planning report stage, the engineering geological conditions of the construction sites of each special item shall be ascertained, the main engineering geological problems shall be assessed, and the geological data and suggestions required for the design of foundation and slope treatment of each structure of the recommended scheme shall be provided.

4   At the detailed design stage, construction geology shall be addressed, the engineering geological conclusions of preliminary investigation shall be reviewed and verified during construction excavation, the geological data required for detailed design shall be provided, and supplementary investigation should be carried out to address new major engineering geological problems.

3.0.4   The investigation stages of small-sized special items of a reservoir should be simplified or combined.

3.0.5   The investigation of the railway, road, water transport, electric power and water resources projects involved in the reservoir special items shall comply with the relevant sector standards. Their sites and surroundings shall, as far as possible, avoid adverse geological effects, such as reservoir landslides, bank collapse, bank deformation, immersion, waterlogging, and goaf collapse; if there still exist some adverse geological effects, geological investigation shall be conducted accordingly, and its level of detail shall meet the requirements of geological investigation of the special item at each stage.

3.0.6   The investigation of reservoir bank protection works involved in the special reservoir items shall comply with the current sector standards NB/T 10138, *Specification for Engineering Geological Investigation of Reservoir Bank Protection for Hydropower Projects*.

3.0.7 The engineering geological assessment of reservoir special items shall be carried out considering the identification results of impoundment-affected areas.

3.0.8 A geological investigation report or geological description shall be prepared at each stage, and the report shall contain the main body, attached drawings, and attachments.

4 Investigation of Reservoir Special Items

4.1 Investigation of Large- and Medium-Sized Special Items

4.1.1 The investigation content, methods, assessment and reports of railway projects shall not only comply with the current sector standard TB 10012, *Code for Geology Investigation of Railway Engineering*, but also meet the following requirements:

1. The reservoir adverse geological phenomena in the project area shall be ascertained.

2. The routes of railways and the sites of bridges, tunnels and stations along the railways should avoid the sections affected by predicted reservoir adverse geological phenomena. If unavoidable, special investigation shall be carried out, and appropriate monitoring may be arranged.

3. Various reservoir adverse geological processes shall be assessed considering the identification results of the impoundment-affected area, their impacts on the project and the hazard degree shall be analyzed, and treatment recommendations shall be provided.

4.1.2 The investigation content, methods, assessment and reports of road projects shall not only comply with the current sector standard JTG C20, *Code for Highway Engineering Geological Investigation*, but also meet the following requirements:

1. The reservoir adverse geological phenomena in the project area shall be ascertained.

2. The routes of roads and the sites of bridges and tunnels on the roads should avoid the sections affected by predicted reservoir adverse geology such as reservoir landslides and bank collapse. If unavoidable, special investigation shall be carried out and appropriate monitoring may be arranged.

3. Various reservoir adverse geological processes shall be assessed considering the identification results of the impoundment-affected area, their impacts on the project and the hazard degree shall be analyzed, and treatment recommendations shall be provided.

4.1.3 The investigation content, methods, assessment and reports of large passenger and freight wharves and ferry wharves shall not only comply with the current sector standard JTS 133, *Code for Geotechnical Investigation on*

Port and Waterway Engineering, but also meet the following requirements:

1. The overall stability of the waterfront bank slopes at the project site and the reservoir adverse geological phenomena around the project site shall be ascertained.

2. The project site shall avoid the sections affected by reservoir adverse geological phenomena and shall be arranged on the stable bank slope section. If unavoidable, special investigation shall be carried out and appropriate monitoring may be arranged.

3. The site suitability and the slope and foundation stability before and after reservoir impoundment shall be assessed considering the identification results of the impoundment-affected area, and treatment recommendations shall be provided.

4.1.4 The investigation content, methods, assessment and investigation results of electric power projects of 110 kV and above shall comply with the current standards of China GB 50548, *Code for Investigation and Surveying of 330 kV ~ 750 kV Overhead Transmission Line*; DL/T 5049, *Technical Code for Exploration and Surveying of Large Crossing Overhead Transmission Line*; DL/T 5076, *Technical Code of Exploration and Surveying for 220 kV and Lower Level Overhead Transmission Line*; DL/T 5122, *Technical Code of Exploration and Surveying for 500 kV Overhead Transmission Line*; and DL/T 5170, *Technical Code for Investigation of Geotechnical Engineering of Substation*. In addition, the following requirements shall be met:

1. The reservoir adverse geological phenomena in the project area shall be ascertained.

2. The sites of substations and transmission towers shall avoid the sections affected by predicted adverse geology such as reservoir landslides and bank collapse. The borehole depth shall be increased at the substation or tower site affected by reservoir adverse geological processes.

3. Various reservoir adverse geological processes shall be assessed considering the identification results of the impoundment-affected area, their impacts on the project and the hazard degree shall be analyzed, and recommendations on engineering measures shall be provided.

4.1.5 The investigation content, methods, assessment and results of water resources projects shall comply with the current standards of China GB 50487, *Code for Engineering Geological Investigation of Water Resources and Hydropower*; and SL 55, *Specification of Engineering Geological Investigation for Medium-Small Water Conservancy and Hydropower Development*. In addition, the following requirements shall be met:

1. The reservoir adverse geological phenomena in the project area shall be ascertained.

2. The canals, pipelines, pumping stations, tunnels, bridges and aqueducts shall avoid the sections affected by the predicted adverse geology such as reservoir landslides, bank collapse, and goaf collapse. The borehole depth shall be increased at the pumping station site affected by the adverse geological processes.

3. Various reservoir adverse geological processes shall be assessed considering the identification results of the impoundment-affected area, their impacts on the project and the hazard degree shall be analyzed, and the recommendations on engineering measures shall be provided.

4.2 Investigation of Small-Sized Special Items

4.2.1 The investigation content, assessment and reports of automobile service roads and tractor roads should comply with the current sector standards JTG C20, *Code for Highway Engineering Geological Investigation*; and DL/T 5379, *Specification of Planning and Designing for Resettlement Special Item for Hydroelectric Project*. The investigation may be simplified appropriately according to the design and site conditions. The investigation of large and extra-large bridges shall be in accordance with Article 4.1.2 of this specification. The investigation methods for automobile service roads and tractor roads shall meet the following requirements:

1. At the resettlement planning outline stage, the investigation should focus on engineering geological mapping, and the mapping scale should be 1 : 10 000 to 1 : 50 000.

2. At the resettlement planning report stage, the engineering geological mapping scale of the route area should be 1 : 2 000 to 1 : 10 000, and the scale of the bridge area should be 1 : 500 to 1 : 2 000. The engineering geological mapping should be carried out along the route area and should cover no less than 100 m on either side of the road, which may be appropriately extended where adverse geology is encountered.

3. At the detailed design stage, the scale of supplementary engineering geological mapping should be 1 : 2 000.

4.2.2 The investigation content, methods, assessment and reports of general passenger and freight wharves and ferry wharves should comply with the current sector standard JTS 133, *Code for Geotechnical Investigation on Port and Waterway Engineering*, and the sites should avoid the sections affected by reservoir adverse geology. The investigation may be simplified appropriately according to the design and site conditions. The investigation methods and geological assessment shall meet the following requirements:

1   At the resettlement planning outline stage, the investigation should focus on engineering geological mapping, and the scale should be 1 : 2 000 to 1 : 5 000.

2   At the resettlement planning report stage, the scale of engineering geological mapping should be 1 : 500 to 1 : 2 000. Where adverse geology affects the site stability or the geological conditions are complex, appropriate exploration and tests should be arranged.

3   At the detailed design stage, the scale of supplementary engineering geological mapping should be 1 : 500 to 1 : 1 000.

4   Geological assessment shall analyze the site suitability and the slope and foundation stability before and after reservoir impoundment considering the identification results of the impoundment-affected area, and recommendations on engineering measures shall be provided.

4.2.3   The investigation content, methods, assessment and reports of overhead transmission line below 110 kV and substation projects should comply with the current sector standards DL/T 5076, *Technical Code of Exploration and Surveying for 220kV and Lower Level Overhead Transmission Line*; and DL/T 5170, *Technical Code for Investigation of Geotechnical Engineering of Substation*. The sites of substations and towers should avoid the sections affected by reservoir adverse geology. The investigation may be simplified appropriately according to the design and site conditions. The investigation methods and geological assessment shall meet the following requirements:

1   At the resettlement planning outline stage, the investigation should focus on engineering geological reconnaissance. The available data such as aerial (satellite) photographs and remote sensing images may be used to interpret the adverse geology.

2   At the resettlement planning report stage, the engineering geological mapping scale should be 1 : 10 000 to 1 : 50 000. In the case of complex geological conditions, the scale may be appropriately enlarged, and appropriate exploration and tests should be arranged.

3   At the detailed design stage, supplementary engineering geological mapping should be carried out as needed.

4   In geological assessment, various reservoir adverse geological processes shall be assessed considering the identification results of the impoundment-affected area. Their impacts on the substations and towers and the hazard degree shall be analyzed.

4.2.4 The investigation content, methods, assessment and reports of small water resources projects shall comply with the current sector standards SL 55, *Specification of Engineering Geological Investigation for Medium-Small Water Conservancy and Hydropower Development*; and DL/T 5379, *Specification of Planning and Designing for Resettlement Special Item for Hydroelectric Project*. In addition, the following requirements shall be met:

1   The canals, main structures on canals and pumping stations should avoid the sections affected by reservoir adverse geology.

2   Various reservoir adverse geological processes shall be assessed considering the identification results of the impoundment-affected area, and their impacts on the canal routes and pumping stations shall be comprehensively analyzed.

5 Assessment of Reservoir Adverse Geological Phenomena

5.1 Landslide

5.1.1 The investigation content and methods shall comply with the current standards of China GB 50287, *Code for Hydropower Engineering Geological Investigation*; and DL/T 5337, *Technical Code for Engineering Geological Investigation of Slope for Hydropower and Water Resources Project*.

5.1.2 Geological assessment shall meet the following requirements:

1 Study the cause, development pattern, type, and development stage of landslides based on the data about topography and geomorphy, stratigraphy and lithology, geological structure, physical weathering, hydrogeology, parameters of rocks and soils, signs of deformation and failure, etc.

2 Perform the stability analysis, predict the stability change trend, assess the adverse effects of landslides on the reservoir special items, and propose treatment recommendations, according to the investigation and monitoring results.

3 Assess the hazards of landslide surge to the reservoir special items according to the location, type, stability and development trend of the landslide.

5.2 Reservoir Bank Collapse

5.2.1 The investigation content and methods shall comply with the current standards of China GB 50287, *Code for Hydropower Engineering Geological Investigation*; and NB/T 10131, *Specification for Reservoir Area Engineering Geological Investigation of Hydropower Projects*.

5.2.2 Geological assessment shall meet the following requirements:

1 For the prediction of reservoir bank collapse, the engineering geological analogy method, graphical method and calculation method shall be adopted, and at least two methods should be used for comprehensive analysis and determination.

2 For the prediction of reservoir bank collapse, the profiles spaced 50 m to 100 m apart should be used to predict the final width of the reservoir bank collapse after impoundment, and determine the extent of the bank collapse-affected area.

3 The impact of reservoir bank collapse on the site stability and suitability of the reservoir special items shall be assessed.

4 The adverse impact of reservoir bank collapse on the reservoir special items shall be assessed, and treatment recommendations shall be proposed.

5.3 Reservoir Bank Deformation

5.3.1 The investigation content and methods shall comply with the current sector standard NB/T 10129, *Specification for Preparation of Special Geological Report on Impoundment-Affected Area for Hydropower Projects*.

5.3.2 Geological assessment shall meet the following requirements:

1 Study the deformation cause, failure type, and deformation stage of the deformed reservoir bank based on the data about topography and geomorphy, stratigraphy and lithology, geological structure, physical weathering, hydrogeology, parameters of rocks and soils, signs of deformation and failure, etc.

2 Perform the stability analysis of the deformed bank, predict the stability change trend, assess the adverse impact of the deformed bank on the reservoir special items, and propose treatment recommendations, according to the investigation and monitoring results.

5.4 Reservoir Immersion

5.4.1 The investigation content and methods and the reservoir immersion prediction methods shall comply with the current standards of China GB 50287, *Code for Hydropower Engineering Geological Investigation*; and NB/T 10131, *Specification for Reservoir Area Engineering Geological Investigation of Hydropower Projects*.

5.4.2 Geological assessment shall meet the following requirements:

1 Determine the boundary conditions of the phreatic aquifer, the burial conditions and distribution of the aquiclude or bedrock, and the parameters for reservoir immersion prediction.

2 Predict the extent of the immersion area, assess the adverse impact of reservoir immersion on the subsoils and foundations of the reservoir special items, and propose recommendations on engineering measures.

5.5 Waterlogging

5.5.1 The investigation content and methods shall comply with the current standards of China GB 50287, *Code for Hydropower Engineering Geological*

Investigation; and NB/T 10131, *Specification for Reservoir Area Engineering Geological Investigation of Hydropower Projects*.

5.5.2 Geological assessment shall meet the following requirements:

1 Analyze the development characteristics and distribution pattern of karst caves.

2 Analyze the impact of underground backwater on the discharge of karst basins, depressions, and valleys after reservoir impoundment, as well as the influence on the underground river discharge due to its outlet being submerged, and predict the waterlogging-affected area.

3 Assess the impact of reservoir waterlogging on the sites and foundations of reservoir special items, and propose recommendations on engineering measures.

5.6 Goaf Collapse

5.6.1 The investigation content shall meet the following requirements:

1 Ascertain the topography and geomorphy, stratigraphy and lithology, geological structure, physical-geological phenomena, and hydrogeological conditions.

2 Ascertain the distribution of goafs and the characteristics of surface deformation.

5.6.2 The investigation methods shall meet the following requirements:

1 The relevant data on mineral mining and goaf distribution should be collected.

2 The methods of engineering geological surveying and mapping should be adopted. The mapping scale may be 1 : 1 000 to 1 : 2 000. Where geological conditions are complex, a small amount of geophysical exploration and drilling exploration may be arranged.

3 The geophysical exploration should be determined comprehensively according to the topographical and geological conditions of the site, the extent of goaf collapse, and the buried depths of goafs. The effective detection area shall be larger than the proposed site to a certain extent, and at least two geophysical exploration profiles should be arranged.

4 The layout of boreholes shall be determined comprehensively according to the collected data, the geophysical exploration results , and the possible influence of goafs. The boreholes shall penetrate the goaf floor no less than 3 m.

5.6.3 Geological assessment shall meet the following requirements:

1. Carry out the stability analysis and assessment of goaf collapse under the action of reservoir water.

2. Assess the impact of earthquakes on the goaf collapse for the regions with a seismic precautionary intensity of Ⅶ or above.

3. Assess the suitability of proposed construction sites.

4. Assess the impact of goaf collapse on structures and propose treatment recommendations.

Explanation of Wording in This Specification

1. Words used for different degrees of strictness are explained as follows in order to mark the differences in executing the requirements in this specification:

    1) Words denoting a very strict or mandatory requirement:

       "Must" is used for affirmation; "must not" for negation.

    2) Words denoting a strict requirement under normal condition:

       "Shall" is used for affirmation; "shall not" for negation.

    3) Words denoting a permission of a slight choice or in an indication of the most suitable choice when conditions permit:

       "Should" is used for affirmation; "should not" for negation.

    4) "May" is used to express the option available, sometimes with the conditional permit.

2. "Shall meet the requirements of…" or "shall comply with…" is used in this specification to indicate that it is necessary to comply with the requirements stipulated in other relative standards and codes.

List of Quoted Standards

GB 50287,	*Code for Hydropower Engineering Geological Investigation*
GB 50487,	*Code for Engineering Geological Investigation of Water Resources and Hydropower*
GB 50548,	*Code for Investigation and Surveying of 330 kV ～ 750 kV Overhead Transmission Line*
NB/T 10129,	*Specification for Preparation of Special Geological Report on Impoundment-Affected Area for Hydropower Projects*
NB/T 10131,	*Specification for Reservoir Area Engineering Geological Investigation of Hydropower Projects*
NB/T 10138,	*Specification for Engineering Geological Investigation of Reservoir Bank Protection for Hydropower Projects*
DL/T 5049,	*Technical Code for Exploration and Surveying of Large Crossing Overhead Transmission Line*
DL/T 5076,	*Technical Code of Exploration and Surveying for 220 kV and Lower Level Overhead Transmission Line*
DL/T 5122,	*Technical Code of Exploration and Surveying for 500 kV Overhead Transmission Line*
DL/T 5170,	*Technical Code for Investigation of Geotechnical Engineering of Substation*
DL/T 5337,	*Technical Code for Engineering Geological Investigation of Slope for Hydropower and Water Resources Project*
DL/T 5379,	*Specification of Planning and Designing for Resettlement Special Item for Hydroelectric Project*
TB 10012,	*Code for Geology Investigation of Railway Engineering*
JTG C20,	*Code for Highway Engineering Geological Investigation*
JTS 133,	*Code for Geotechnical Investigation on Port and Waterway Engineering*
SL 55,	*Specification of Engineering Geological Investigation for Medium-Small Water Conservancy and Hydropower Development*